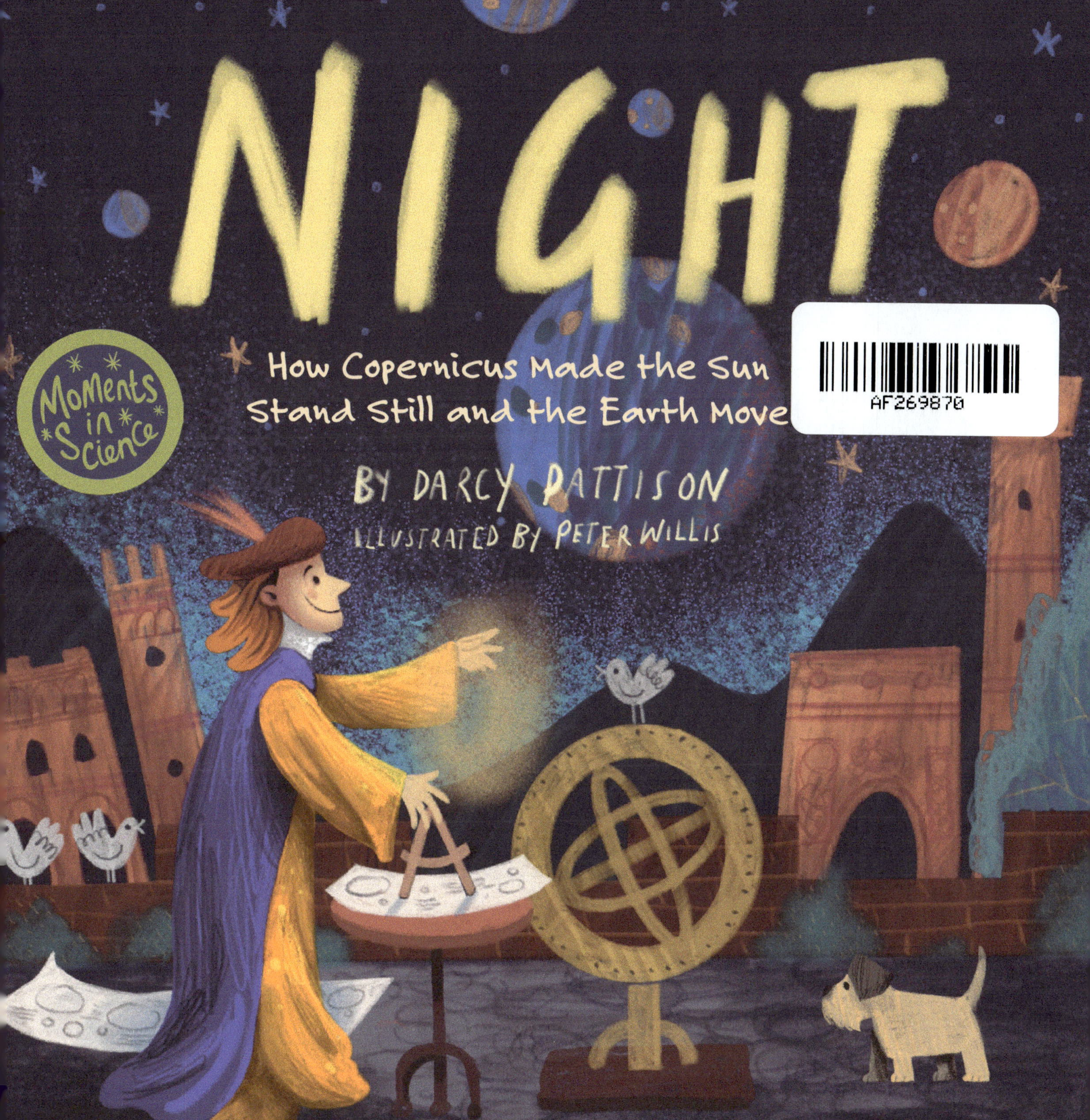

NIGHT
How Copernicus Made the Sun Stand Still and the Earth Move
BY DARCY PATTISON
ILLUSTRATED BY PETER WILLIS
Moments in Science

Mims House
1309 Broadway
Little Rock, AR 72202

MimsHouseBooks.com

Publisher's Cataloging-in-Publication Data

Names: Pattison, Darcy, author. | Willis, Peter,
illustrator.
Title: Night : how Copernicus made the sun stand
still and the earth move / by Darcy Pattison; illustrated
by Peter Willis.
Series: Moments in Science
Description: Little Rock, AR: Mims House, 2026.
| Summary: In the 16th century, a curious Copernicus
challenges old ideas and theorizes that Earth moves
around the Sun, changing how humanity understands
the solar system, science, and our place in the
universe.
Identifiers: LCCN: 2026900525| ISBN:
9781629443355 (hardcover) | 9781629443362
(paperback) | 9781629443379 (ebook) | 9781629443386
(audio)
Subjects: LCSH Copernicus, Nicolaus,
1473-1543--Juvenile literature. | Astronomers--Poland--
Biography--Juvenile literature. | Astronomy--History--
Juvenile literature. | BISAC JUVENILE NONFICTION
/ Biography & Autobiography / Science & Technology
| JUVENILE NONFICTION / Science & Nature /
Astronomy | JUVENILE NONFICTION / Science &
Nature / History of Science
Classification: LCC QB36.C8 P38 2026 | DDC 520.92-
-dc23

Look up! Jumbled stars glitter in the night sky. But watch the sky long enough and you'll start to see the patterns of how the moon, planets, and stars move across the sky.

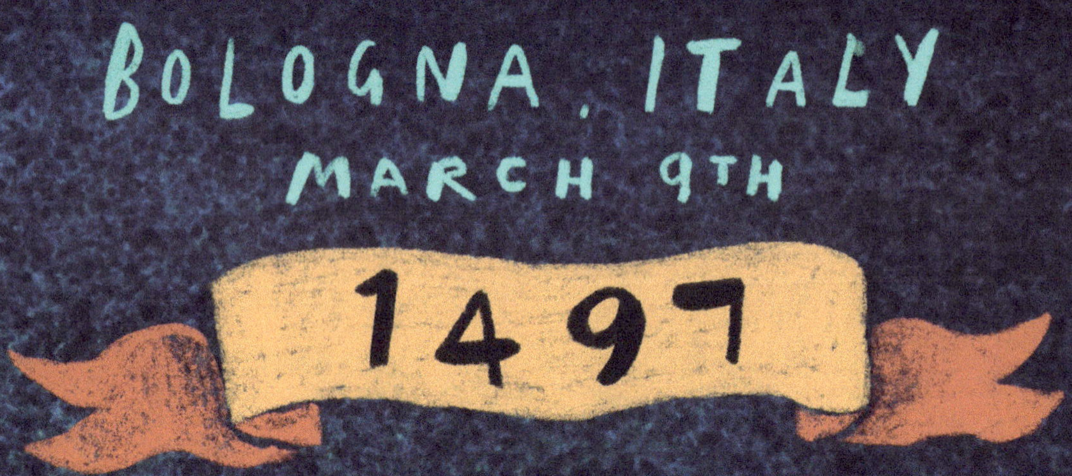

24-year-old Nicholas Copernicus studied the slim crescent moon over Bologna, Italy. His job that night was to assist the astronomer Domenico Maria da Novara. At this time, there were no telescopes, so astronomers used only their eyes. They planned to watch the moon cross in front of the star Aldebaran—the eye of the bull—in the Taurus constellation. For a time, the moon would hide Aldebaran's twinkle.

But Nicholas and Domenico were mathematicians, not just astronomers. They wanted to measure how long the star was hidden by the moon.

WHY?

The science of the time,

which followed the theory developed by Ptolemy (TALL-uh-me), said that Earth was the center of the universe. If the moon, planets, and stars moved around Earth, as Ptolemy's theory said, Aldebaran would be hidden for a long time.

As the Moon glided across the sky.

the star disappeared. Nicholas timed it and watched to see when Aldebaran would reappear.

Oh! Aldebaran was shining again, but far too early.
Ptolemy's theory was wrong about the moon's movement.
Nicholas looked at observations from other astronomers,
and the math almost never worked as predicted.

Domenico and Nicholas talked about Ptolemy's theory.
If the math was wrong here, what else might be wrong?
And how else could they explain the patterns of objects
moving across the heavens at night? Nicholas decided
to study these patterns, as an astronomer and a
mathematician.

Nicholas finished his studies in church law and medicine and started working for the Catholic Church as a cleric, or clergyman. In 1510, Nicholas moved to Frombork, Poland, to work at the cathedral that rose above the Baltic Sea, a quiet place with dark skies. Happily, he climbed the stairs to his new observatory in the town hall tower. Here, he would work on his theories for the next thirty-three years.

As a cleric, Nicholas often traveled, and sometimes he could talk with other scientists about his astronomy observations.

Some scientists thought the center of the universe had a divine spark, and that meant it was the sun.

Older theories also put the sun at the center. Nicholas measured the night sky again.

The idea grew: what if he made the sun stand still at the center of the universe and put Earth in motion around the sun?

During his night observations, Nicholas used simple instruments to measure how far each planet was from the sun. Venus and Mercury were close to the sun, which made them inner planets. That meant they were closer to the sun than Earth. Mars was farther from the sun than Earth, making it an outer planet.

Armillary Sphere

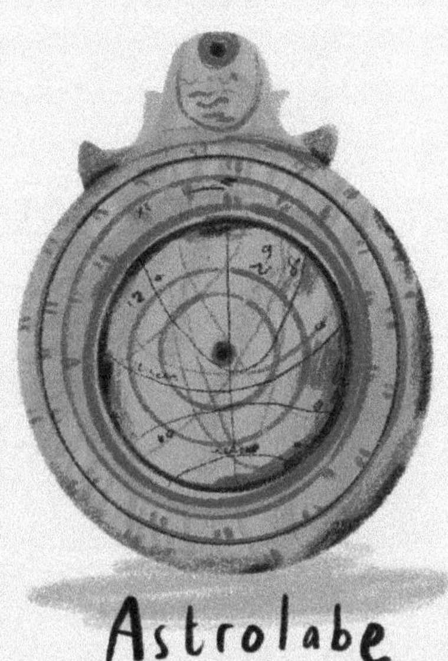

Astrolabe

Qvadrant
Triquetrum

    Nicholas put the planets in order from the sun as Mercury, Venus, Earth, Mars, Jupiter, and Saturn. Without telescopes, he couldn't see planets farther out than Saturn.

Earth
Mars
Jupiter
Saturn
Venus

With the planets in order, Nicholas had
a new theory about the universe.

He kept Ptolemy's idea that heavenly
bodies moved in circles around something
in the center. But Nicholas put the sun in
the center, a theory known as heliocentric,
because *helio* is Greek for "sun" and *centric*
means "in the center."

He sent Earth
spinning around the sun.

Using the new theory, Nicholas predicted when and where different planets could be observed. Over and over, he watched the starry sky. The planets gleamed and glittered exactly where he expected. The math worked.

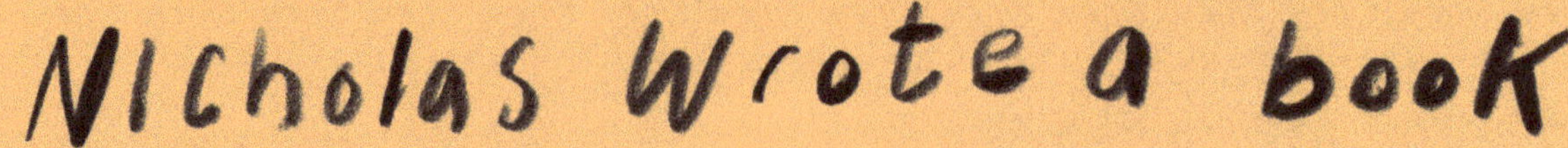

Nicholas wrote a book

about his new heliocentric theory. He only told a few people about his book, though, because he thought the math wasn't quite right yet. However, Pope Clement VII read Nicholas's theory in 1533 and was excited by the new ideas.

In 1543, On the Revolutions of the Heavenly Spheres
(De revolutionibus orbium coelestium) was published.
History says that Nicholas held a copy of his book
on his deathbed.

For years, some scientists found Nicholas's ideas interesting, while others thought they were wrong. Then, 73 years after his death, in 1616, Nicholas's book was banned by the Catholic Church. Pope Paul V believed that the center of the universe was Earth, not the sun. Some scientists also worried that Nicholas's theory would challenge other scientific teachings.

219 years later.

in 1835, the Catholic Church removed the book ban because of the work of other astronomers.

Today, We Understand

that Nicholas's theory and math were basically correct. The sun is the center of our solar system, not Earth.

Nicholas Copernicus will always be remembered as the man who made the sun stand still and set the Earth in motion.

NICHOLAS COPERNICUS
(February 19, 1473–May 24, 1543)

Nicholas Copernicus was born in Toruń, Royal Prussia, now known as Poland, to German-speaking parents. Nicholas was the youngest of four and had two brothers and one sister. His father, a copper merchant, died when Nicholas was ten years old. His uncle Lucas Watzenrode the Younger (1447–1512) made sure his nephews were educated and found a career. Nicholas studied at the University of Krakow, in Poland, and at the University of Bologna and the University of Padua, both in Italy. He then returned to Poland and worked at the Frombork Cathedral for the rest of his life.

Because he worked for the Catholic Church, Nicholas didn't have a regular schedule of making observations, and many of his papers are lost. Historians estimate that he made over 100 observations of the night skies. He concentrated on his heliocentric theory (that planets revolve around the sun), working out the math and making observations to confirm his theory.

Nicholas used only basic observation tools. The triquetrum, a three-cornered tool (tri means three), measured the altitude of stars, or how high they were in the sky. He also used a quadrant, a quarter circle, to help measure the stars' altitude. Then he used an astrolabe, a complicated set of disks, to calculate the math of the observations.

About 65 years after Nicholas died, the telescope was invented. Galileo Galilei (1564–1642) improved the telescope so that it made things 30 times bigger. His observations of Venus brought even more support for Copernicus's ideas.

SOURCES

On the Revolutions of the Heavenly Spheres, by Nicholas Copernicus, translated by Charles Glen Wallis, St. John's Bookstore, Annapolis, 1939. Originally published as *De revolutionibus orbium coelestium*, by Nicolai Copernici Torinensis. Johanes Petreius, Nuremberg, 1543.

In 2023, the author visited three sites in Poland: University of Krakow, in Krakow; the Nicholas Copernicus House, in Toruń; and Frombork Cathedral, in Frombork. Thanks to the Magda Listos family for their hospitality while the author was in Poland.

WHAT IS A SOLAR SYSTEM?

When Nicholas Copernicus looked at the night sky, he saw many stars, some larger than others. He knew that some stars appeared in the night sky in a regular pattern across the year. These were the known planets: Mercury, Venus, Mars, Jupiter, and Saturn. Of course, Earth is a planet, too, but Copernicus was standing on Earth, so he couldn't see it in the sky. Without a telescope, he couldn't see Uranus or Neptune either.

While Copernicus realized that the planets circled the sun, he didn't understand why. Later astronomers, including Johannes Kepler, Galileo Galilei, and Isaac Newton, discovered gravity. Gravity is the force that pulls two objects toward each other. The sun is larger than the planets, so it has the biggest gravity, or the biggest pull. But the planets are moving sideways, so instead of falling into the sun, they circle it. A solar system is composed of a sun, plus everything caught in the sun's gravity: planets, moons, asteroids, and space dust.

PLANETS IN OUR SOLAR SYSTEM

At the center of our solar system is the sun, known as Sol, a medium-sized star. Over 1.3 million Earths could fit into the sun.

Mercury is the planet closest to the sun, and it's the smallest planet. A year (the time to circle the sun) on Mercury is just 88 Earth days.

Venus is the hottest planet because its atmosphere holds in the heat. It rotates backward from Earth, so the sun rises in the west.

Earth has so much water that it covers about 71% of the planet's surface. It's the only known place in the solar system that supports life. It has one moon.

Mars has two moons, Phobos and Deimos. It's sometimes called the Red Planet, a name that comes from the iron-rich soil that looks like rust (iron oxide). A Martian day lasts 24 hours, 37 minutes, just a bit longer than an Earth day.

Jupiter is the largest planet and has at least 95 moons. The Great Red Spot, a gigantic storm, has been observed on Jupiter's surface for over 400 years.

Saturn is famous for its spectacular ring system, which is mostly made of ice particles. A year on Saturn lasts 29 Earth years. Each season—spring, summer, fall, or winter—lasts over seven Earth years.

Uranus is tilted on its side, so as it circles or orbits the sun, it rolls like a ball. It was the first planet discovered by using a telescope. A year on Uranus takes 84 Earth years.

Neptune is the windiest planet. It's the first planet found by mathematical prediction rather than observation. A year on Neptune is about 165 Earth years.

Pluto was once thought to be a planet too. Today, scientists consider it a dwarf planet because while it circles the sun, it doesn't have strong enough gravity to push or pull away other objects in its path.

FROMBORK CATHEDRAL'S DARK SKIES

For most of Nicholas Copernicus's life, he lived and worked at Frombork Cathedral, which lies on the Baltic Sea. The cathedral was remote, only reached by difficult travel. The night skies were dark, a perfect place for an astronomer.

Today, light pollution means our nights are artificially lit, changing the night into a semi-dark time of day. Too much light can disrupt the lives of nocturnal (active at night) animals, change migration and mating behaviors, and impact food webs. For example, light pollution can affect the mating of fireflies, and migrating birds can lose their way or hit buildings. It also disconnects people from the rhythms of nature.

DarkSky International recognizes, protects, and encourages dark skies. Check to see if you are close to a location designated as an International Dark Sky Place at darksky.org.

www.ingramcontent.com/pod-product-compliance
Lightning Source LLC
LaVergne TN
LVHW071912090626
840870LV00043B/725